WHAT IS SELENOLOGY?

Selenology is the study of Earth's Moon.

The scientists who study the Moon are called **SELENOLOGISTS.**

WILL WE EVER LIVE ON THE MOON?

DISCOVER THE SCIENCE BEHIND **SELENOLOGY**

(seh-luh-NOH-luh-jee)

Written by Olivia Watson

Illustrated by Verónika Cháves Morales

Words that are tricky to understand are in **bold**. Find out what they mean in the glossary.

Words that are difficult to say are in *italics*. Find out how to say them at the back of the book.

Humans have always gazed at the Moon shining in the sky and wondered what it's like. What's up there? What's it made of? And, the biggest question of all, could we live there someday?

For thousands of years, people have watched the Moon move across the sky, appearing to change shape day by day.

Long ago, humans made basic calendars to track time using the **phases** of the Moon. Today, we understand that the Moon doesn't really change shape – it's a wonderful trick of

SUNLIGHT AND SHADOWS!

Everything changed when the **telescope** was invented just over 400 years ago! **Astronomers** like *Galileo Galilei* zoomed in on the Moon and saw that it was covered in mountains and **craters**.

Selenology became a special part of the science of *astronomy*. Scientists called *selenologists* studied the Moon in more and more detail.

But selenologists could only learn so much from here on Earth. In the twentieth century, engineers started to build amazing machines like rockets that could help get an even closer look at the Moon.

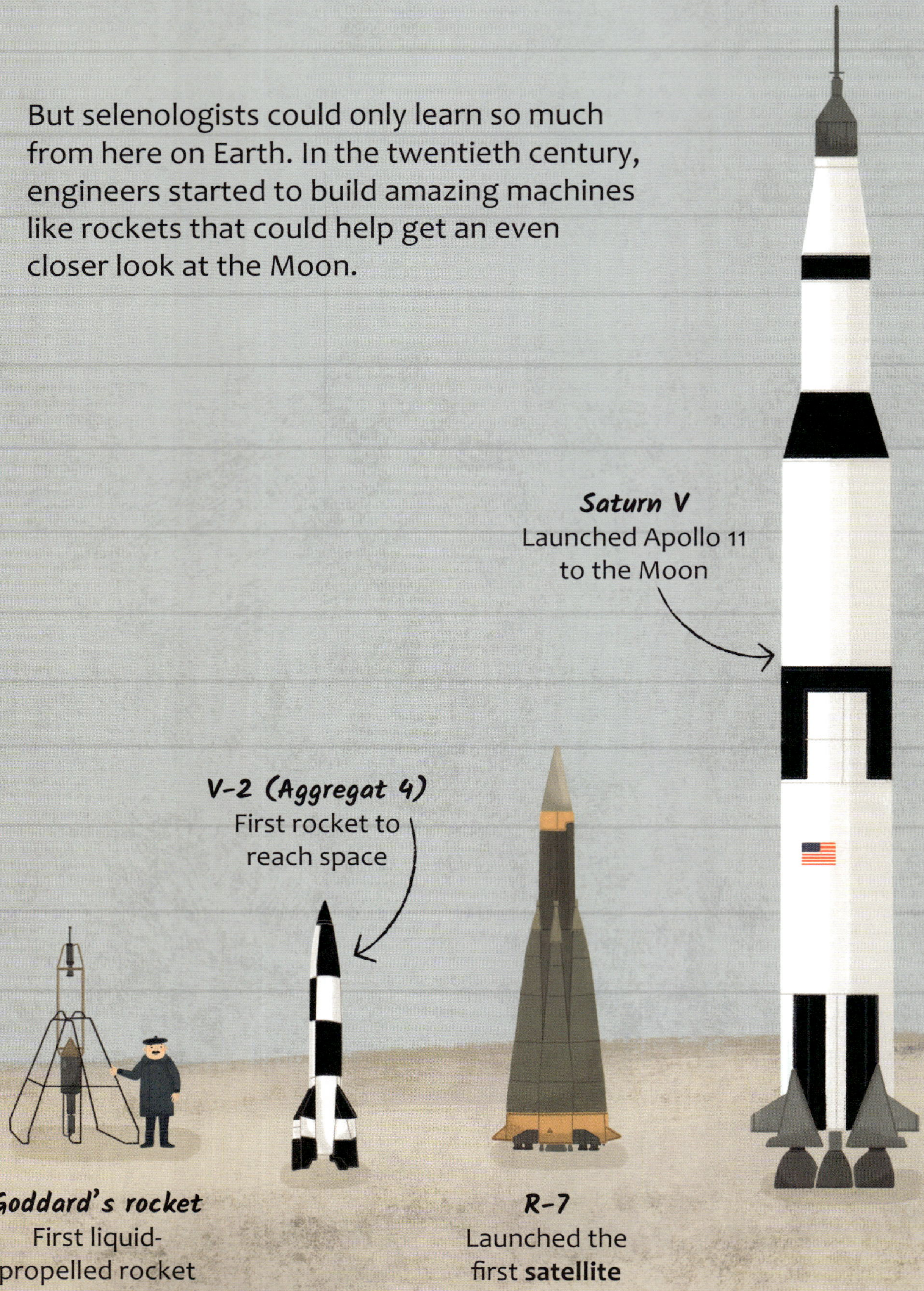

The first rockets were small and didn't fly far, but they got bigger and stronger until eventually they were able to

VISIT THE MOON AND BEYOND!

Space shuttle
First reusable spacecraft to be built

Atlas V-541
Launched robots to Mars

The journey took three days, but finally humans landed on the Moon. Walking was different there – the low **gravity** made the **astronauts** bounce around. And the lack of weather like wind meant their footprints weren't blown away!

People had got
to the Moon and
back safely, but could
we really live there?

Thanks to samples collected by the astronauts, selenologists could study Moon dust, soil, and rocks.

It was the first time they could see what the Moon is made of. They discovered the Moon has amazing **minerals** and metals that are hard to find on Earth!

These materials are so useful that many scientists think astronauts should go back to the Moon to collect more!

They're especially interested in a **substance** called Helium-3, which could be turned into a powerful type of energy…

and other **rare** Earth metals that are used in mobile phones, computers, and medical equipment.

Selenologists don't just test rocks! The astronauts set up experiments that sent information back to Earth long after they'd left.

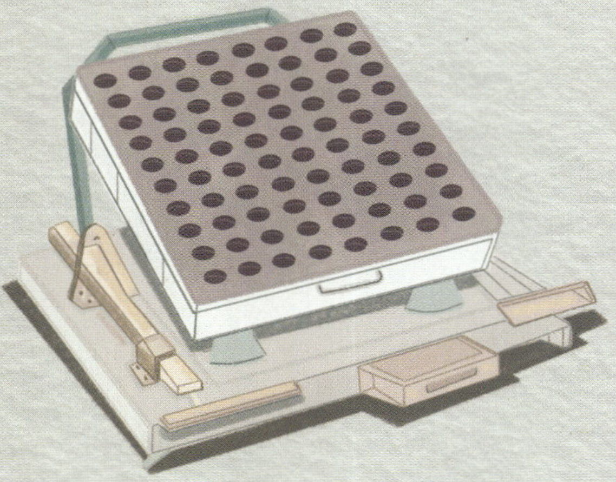

Laser ranging retroreflector

Lunar dust detector

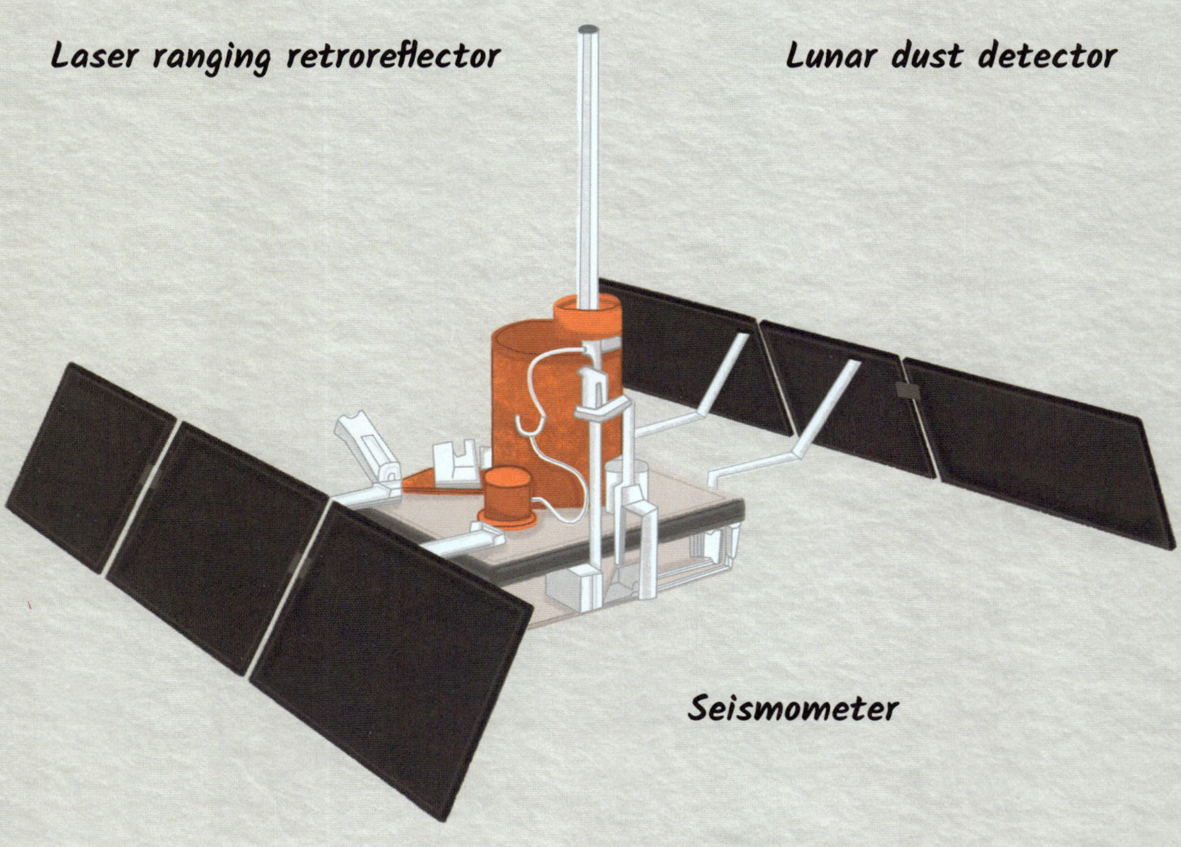

Seismometer

Lunar surface magnetometer

Solar wind spectrometer

Solar wind composition experiment

There are devices that tell us about the Moon's **rotation** and **orbit.** Some sensors can detect **moonquakes.** And there are many other tools to help scientists understand the Moon.

Studying the Moon has even helped scientists learn about the ancient **solar system**! They've discovered that it was once filled with **asteroids** which kept smashing into different worlds. Scientists have already learned a lot from just a few tests, but they're sure the Moon is

HIDING LOTS OF SECRETS!

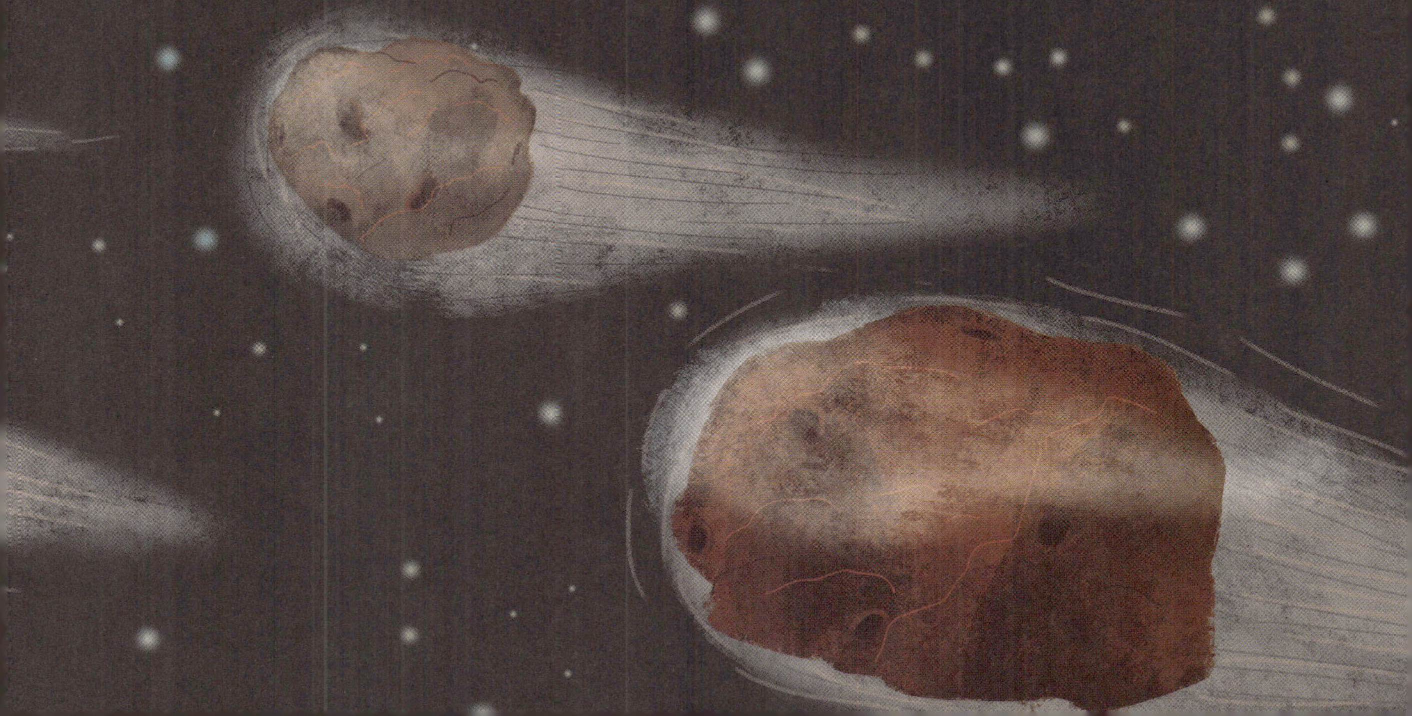

There's a lot of the Moon to explore. It would be better for people to live there rather than make short visits. But there are problems to solve.

Selenologists are helping to discover how we could **generate** energy to last through two weeks of darkness at a time...

whether we could get enough water from Moon ice to survive...

and if we could grow food in the Moon's soil, which is dry and doesn't have many **nutrients**.

There's a lot to figure out, but living on the Moon could be amazing for science! We could set up huge telescopes facing into outer space. There's no **atmosphere** up there to cloud the view, so telescopes would see a lot more than they do from Earth.

The Moon could be the perfect base for exploring outer space! It has much lower gravity than we have on Earth, so it could be easier to launch rockets.

One of **NASA**'s goals is for humans to explore the solar system! It's a long way to travel, but the Moon could be a stepping stone for launching **A MISSION TO MARS.**

NASA wants to set up a base on the Moon's south pole. It would be an incredible place for scientific research – just imagine making amazing discoveries from the Moon!

There are many challenges, but plans are already being made. Before too long, people could indeed be living and working on the Moon!

Journey to the MOON AND BACK

So, one day we might travel to the Moon, but how will we actually get there and back? What do we need to know before we make this big step?

WHERE WILL WE LIVE ON THE MOON?

NASA plans to build a base on the Moon's south pole, where there is plenty of sunlight for **solar power**. But it will probably be quite a few years until a base is built.

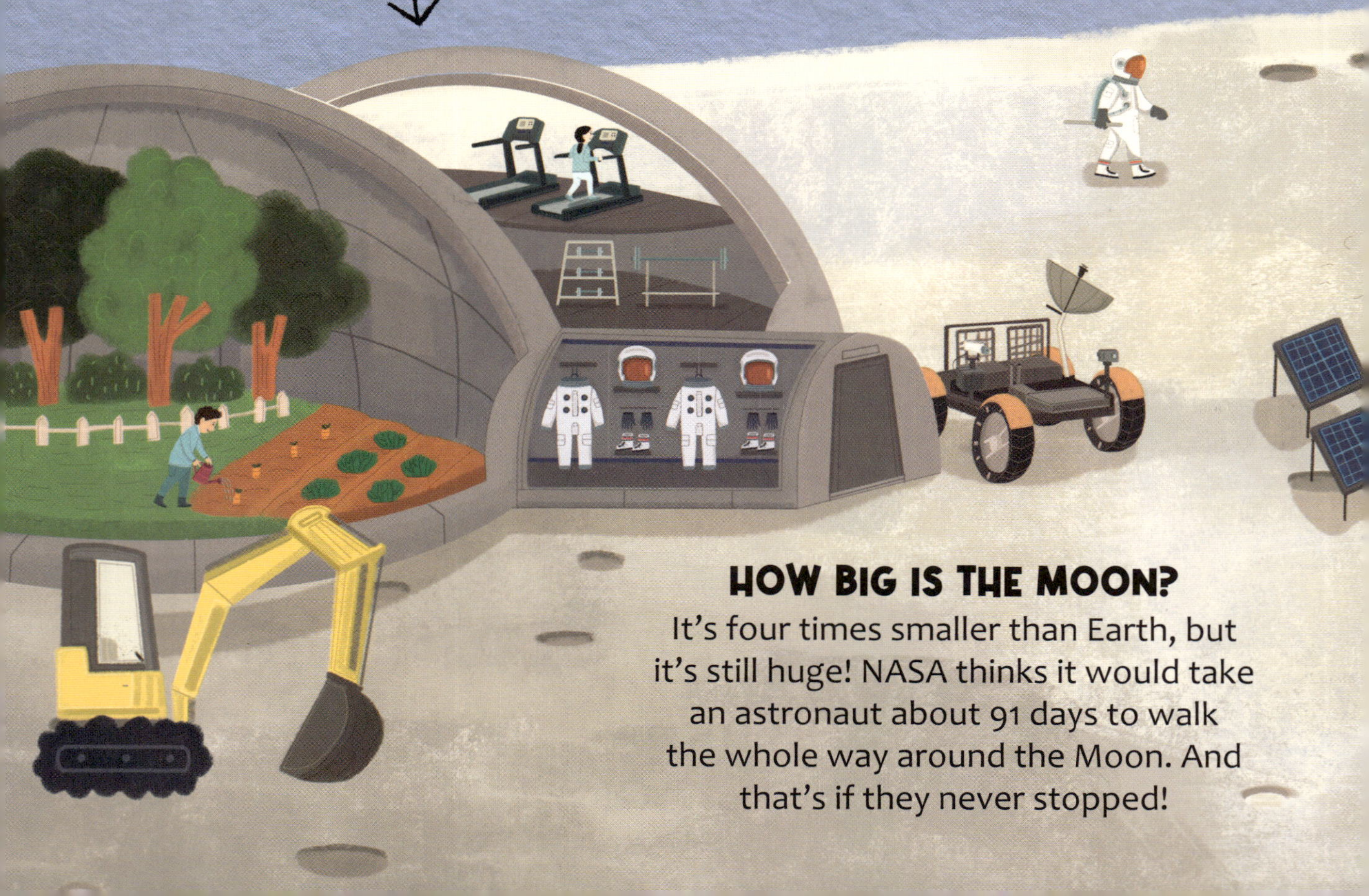

HOW BIG IS THE MOON?

It's four times smaller than Earth, but it's still huge! NASA thinks it would take an astronaut about 91 days to walk the whole way around the Moon. And that's if they never stopped!

WHEN ARE HUMANS GOING BACK TO THE MOON?

Humans haven't been to the Moon for more than 50 years, but many space agencies and companies from around the world aim to go back in the next few years.

HOW LONG DOES IT TAKE TO GET THERE?

It takes at least 3 days to get to the Moon, but it can take much longer depending on the spacecraft, the journey, and the fuel.

HOW WOULD WE TRAVEL ON THE MOON?

Vehicles are quicker than walking! During the **Apollo missions**, Moon buggies were used, but NASA is working on new vehicles for future missions.

Out-of-this WORLD FACTS

There's so much to discover about the world of selenology. Do you know the answers to some of the world's biggest questions about the Moon?

HOW MANY MOONS ARE THERE?

There are hundreds in the solar system. Galileo discovered 4 moons orbiting Jupiter hundreds of years ago. Before then people didn't know any others existed apart from the moon orbiting the Earth. Hence it is called "the moon".

WHAT ACTUALLY IS THE MOON?

Not a planet! Planets orbit the Sun, but the Moon orbits the Earth. This makes it a natural satellite; a space object that orbits a planet.

WHERE DOES MOONLIGHT COME FROM?

The Moon doesn't make its own light. What we call moonlight is really the reflection of sunlight bouncing off the Moon!

WHY DOES THE MOON SPIN AROUND THE EARTH?

The Moon orbits the Earth, in the same way that the Earth orbits the Sun. The smaller space object is kept orbiting the bigger one due to gravity.

DO SELENOLOGISTS ONLY STUDY THE MOON?

Yes! Selenology is a branch of astronomy that only studies Earth's Moon.

GLOSSARY

Apollo missions – a series of space exploration missions led by NASA (see right) in the 1960-70s that saw humans walk on the Moon.

Asteroids – small space rocks that orbit the Sun.

Astronauts – people trained to travel into outer space.

Astronomers – scientists who study astronomy (outer space). *Need help saying this? See below.*

Atmosphere – the gases that surround a planet.

Craters – dents or holes in the surface of something.

Generate – to make.

Gravity – an invisible force that pulls things together. Gravity keeps your feet on the ground and keeps Earth spinning around the Sun.

Minerals – substances that are naturally found in things like rocks, sand, and soil.

Moonquakes – like earthquakes, but on the Moon.

NASA – the National Aeronautics and Space Administration is an agency that deals with space research and exploration.

Nutrients – substances or ingredients that plants and animals need to live and grow.

Orbit – the repeated path taken by one object circling around another object in space.

Phases (of the Moon) – the way the shape of the Moon seems to change over a period of about a month.

Rare – something that is uncommon.

Rotation – the movement of an object turning around a central point.

Satellite – any object that orbits a planet. Satellites can be natural, like moons, or artificial (made by humans), like the ones we use for communication.

Solar power – energy created from the power of the Sun.

Solar system – the Sun and everything that moves around it.

Substance – the stuff or material which something is made of.

Telescope – an instrument that makes faraway objects appear bigger.

HOW DO I SAY?

Astronomers
uh-STROH-nuh-mers

Astronomy
uh-STROH-nuh-me

Galileo Galilei
gah-li-LAY-oh gah-li-LAY

Selenologists
seh-luh-NOH-luh-jists

Selenology
seh-luh-NOH-luh-jee

THE BIG QUESTIONS ANSWERED

This is more than just a series of books; it is a complete resource. Accompanying each book is a variety of FREE material to engage curious kids with science.

www.thebigquestionsanswered.com

Use the QR code to visit the website, download free resources, and discover other books in the series.

On the website, find out incredible things about selenologists, including what they do, some of their greatest discoveries, and the people who have made a difference in this field of science.

The material is also available for home or classroom use, supporting all the information in this book.

Teachers' & Parents' Resources
With discussion prompts and questions, extra information, and facts around key topics.

Young Selenologists' Activity Pack
Fun activities for wannabe Moon explorers, including creative writing, drawing, word searches, and much, much more.

BEETLE BOOKS

The Big Questions Answered is published by Beetle Books. Beetle Books is an imprint of Hungry Tomato Ltd.

First published in 2024 by Hungry Tomato Ltd
F15, Old Bakery Studios, Blewetts Wharf, Malpas Road,
Truro, Cornwall, TR1 1QH, UK.

ISBN 9781835691304

A CIP catalog record for this book is available from the British Library.

With thanks to:
Editor: Holly Thornton
Editor: Millie Burdett
Senior Designer: Amy Harvey
Tim Cook for his valued contribution
The team at Beehive Illustration

Printed and bound in China.

Picture Credits:
(t = top, b = bottom, m = middle, l = left, r = right)
Shutterstock: Ara Hovhannisyan 34bl; Dima Zel 35ml; Dotted Yeti 34mr; Rangizzz 35tl;
True Touch Lifestyle 35bl.